Energy

Use less – save more

100 ENERGY-SAVING TIPS FOR THE HOME

JON CLIFT & AMANDA CUTHBERT

First published in 2006 by
Green Books
Foxhole, Dartington
Totnes, Devon TQ9 6EB
www.greenbooks.co.uk

Reprinted 2007

Design concept by Julie Martin

Printed in the UK by Cambrian Printers on Revive 75,
made from 75% recycled paper (50% post-consumer waste)

DISCLAIMER: The advice in this book is believed to be correct at
the time of printing, but the authors and publishers accept no liability
for actions inspired by this book.

ISBN 978 1 903998 88 5

Contents

Introduction

The cost of electricity has nearly doubled since 2002.

We are all using more and more energy: charging up our mobiles and laptops, keeping our rooms so hot that we walk around in short sleeves in the winter, or drying our clothes in the tumble dryer. But the consequences of using so much so freely is causing our climate to change, and our energy bills to rise.

We're going to have to be more efficient in the way we use this energy. Most of the energy we use comes from fossil fuels, which when burnt to produce the energy we all need, releases vast quantities of CO_2 – the gas that's the main cause of climate change.

Our electricity consumption has gone up 70% since 1970.

The energy we use in our homes comes from oil, gas or electricity. We used to be self-sufficient in oil and gas, but now our reserves in the North Sea are drying up and we increasingly have to rely on foreign sources for our supplies: gas from Russia, and oil from often politically unstable countries. The days of cheap energy are over – reducing our energy consumption is vital.

We can take control of this situation and reduce our energy consumption and our energy bills. We don't have to live shivering in an unheated room with no modern appliances; we're just talking about being more energy-efficient – reducing the need for so much power.

Little things that we can do every day can produce large results. If we all turned off our TVs and other gadgets that are kept on stand-by, for example, we could shut down a couple of power stations in the UK, with huge reductions in CO_2 emissions. Small actions, large results – it's really about being aware, knowing what we can do to have an immediate effect without compromising our quality of life.

> Our homes produce over ¼ of the UK's CO_2 emissions, even more than our cars.

Simple actions can considerably reduce our energy consumption and our energy bills, and help reduce climate change: the less energy we use, the less CO_2 is released, which benefits us all. Once we are actually aware of what's happening, most of the things we need to do are just common sense.

How much electricity do you use?

How much electricity do you use?

The amount of electricity consumed varies hugely according to
which appliance and model you use. Check out the list below
to see which are the hungriest appliances. *All figures given
here are approximate – see your actual appliance for accurate
figures.*

Appliance	Average watts used per hour	Appliance	Average watts used per hour
Low-energy light bulb	11	Iron	1,000
Extractor fan	75	Dishwasher	1,000
Laptop computer	75	Bar heater (one bar)	1,000
Conventional light bulb	100	Washing machine	1,200
Stereo	100	Cooker (one ring)	1,300
Television	100	Oil filled heater	2,000
Video recorder	110	Fan heater	2,000
Refrigerator	125	Bar heater (two bars)	2,000
Desktop computer	150	Deep fryer	2,000
Freezer	300	Cooker oven	2,150
Hair dryer	750	Kettle	2,250
Microwave	750	Immersion heater	3,000
Vacuum cleaner	800	Electric shower unit	8,000
Toaster	1,000	Cooker (everything on)	11,500

Home heating

Home heating

Spend nothing – save money

1. Take control of your heating. Consider turning down the thermostat controlling the temperature of your room or house by 1°C. You will have either a single control at a central position such as in the hall, or thermostats attached to the individual heaters or radiators.

2. Turn radiators off or down in rooms you only use occasionally.

> **WARNING – IF YOU ARE ELDERLY OR INFIRM, TRY TO KEEP YOUR ROOM TEMPERATURES AT LEAST 18°C, AND YOUR LIVING ROOM AND BATHROOM ABOUT 21°C.**

3. You don't necessarily need to turn up the heating for babies: a room temperature of about 16°C–20°C is ideal.

4. Turn down the thermostat when you are going away on holiday: 5°C will prevent pipes bursting in cold weather.

5. Set the timer for your heating system so that it comes on about 30 minutes before you get up, or when you come home in the evening. Switch the heating off about ½ hour before you leave in the morning or go to bed.

6. If you use plug-in electric heaters such as bar heaters, oil filled radiators or panel heaters, use them sparingly as they are very expensive to run.

> **Just lowering the temperature of your thermostat by 1°C can reduce your energy bill by 10%.**

7. Move furniture away from any radiators or heaters, to allow heat to get out into the room.

8. If you are too hot in your room, turn the heating down or off rather than opening a window.

9. Rather than turn up the heat, put on an extra layer of clothes.

10. Draw curtains over windows at night; they provide insulation and help to keep the heat in the room.

11. If your curtains are thin, line them with thicker fluffy materials, such as brushed cotton, to help keep the heat in.

12. Open the curtains during the day if the sun is shining on your windows, and let the sun heat your room.

> Keeping our homes warm during the winter months accounts for about ⅔ of our household energy bills.

13. Avoid covering radiators with curtains – they will funnel the heat out through the glass of the windows. Tuck them in behind, to enable the radiator heat to come into the room.

14. If you do not have double-glazing, you can reduce your heat loss by putting cling film over each window pane. It works very well, will reduce noise coming through the window, and should last the whole of a winter.

15. Keep external doors shut.

Spend a little · save more

> **WARNING** – DON'T BLOCK UP AIR VENTS OR GRILLES IN
> WALLS IF YOU HAVE AN OPEN GAS FIRE, A BOILER WITH AN
> OPEN FLUE, OR A SOLID FUEL FIRE OR HEATER. THESE NEED
> SUFFICIENT VENTILATION TO BURN PROPERLY – OTHERWISE
> HIGHLY POISONOUS CARBON MONOXIDE GAS IS RELEASED.

16. Buy and fit a draught excluder to your letterbox. They only cost a couple of pounds, but make a big difference.

17. Fit draught excluders to external doors and windows. Foam strips are cheap, but if you can afford it, buy the longer-lasting rubber or plastic systems. You may not want to do this in your bathroom or kitchen if you have problems with condensation. *Make sure you still have sufficient ventilation – see above.*

> About ¼ of all the energy we use to heat our homes escapes through single-glazed windows.

18. Stop draughts coming under skirting boards or through floorboards by filling the gaps with strips of wood, cork, or the correct sealant. *Make sure you still have sufficient ventilation – see above.*

19. If your walls are not insulated, put some radiator foil between the radiators and the walls. It's cheap, very effective and easy to install. Actual radiator foil is best; it has a layer of insulation behind the aluminium foil. Ordinary kitchen foil helps, but is less effective. Stick it to the wall with double-sided sticky pads, with the shiny side facing into the room.

20. Insulate your walls. If you have cavity walls, they are easy and quick to insulate, and in most cases it can be done in a day. Solid walls are insulated by placing cladding either inside or outside; it's more complex, but worthwhile, as solid walls lose more heat than cavity walls. *There will probably be a grant available to help you pay for this.*

21. Insulate your loft. This is probably one of the simplest and most effective methods of reducing your heat and energy loss. Loft insulation should be a minimum of 270mm thick. You can do it yourself. There are some very user-friendly materials available, but whichever insulation type you choose, protect yourself with appropriate clothing and a face mask. *There will probably be a grant available to help you pay for the installation.*

22. Fitted carpets with underlay will give you much more insulation than bare boards, and will stop draughts.

23. If you use gas or oil for heating, install a condensing boiler. They are more efficient than conventional boilers, and will save you money and produce less CO_2. *There will probably be a grant available to help you pay for this.*

24. Service your boiler regularly – it will be more efficient and use less energy.

25. Fit a thermostat to every radiator. This will enable you to vary the temperature in different rooms.

Almost 40% of all the heat used to warm rooms escapes through the walls and roof space if they're not insulated.

26. Block up any unused fireplaces to stop heat going up the chimney. You can stuff scrunched-up newspaper into the hole where the fireplace enters the chimney, or buy special balloons to put there. Both these methods are good because they still allow a little air to circulate, which is necessary to prevent damp.

Water heating

Water heating

Spend nothing - save money

27. Turn down the temperature of your hot water at the central heating boiler, at the immersion tank (if your water is heated by electricity), or on your instant water heater. Don't waste energy heating water only to have to add cold water so that it is not too hot to use! 60°C / 140°F should do it.

> Water heated by an electric immersion heater costs more than 30p an hour. Just turn it on half an hour before you need it – and don't forget to turn it off afterwards.

28. You can use less energy by taking a quick shower rather than a bath. If you use a power shower, remember that in five minutes it can use as much energy as a bath.

29. If you want a bath, then why not share it with a friend? It's much more environmentally friendly!

30. Don't leave hot water taps running – use the plug.

31. If your hot water tap is leaking, fix it quickly.

32. If your immersion or hot water tank is not insulated, get an insulating jacket – if not, about three-quarters of the energy you are buying to heat your hot water is wasted. Insulating jackets are not expensive. Buy one that's at least 75mm thick.

33. If you don't have a timer for your immersion heater, buy one. Set it so that the water is only heated when you need it.

> It's a myth that you use less electricity by keeping your immersion heater on all the time. It's much cheaper and consumes less energy if water is heated only when needed.

34. Insulate your hot water pipes. Insulation is cheap and easy to fit: clip it around any uninsulated pipes.

AUTO
OFF
ON
ONCE
ADV.
+ 1 hr.

ON

7:31 PM

ADVANCE   + 1hr.

Lighting

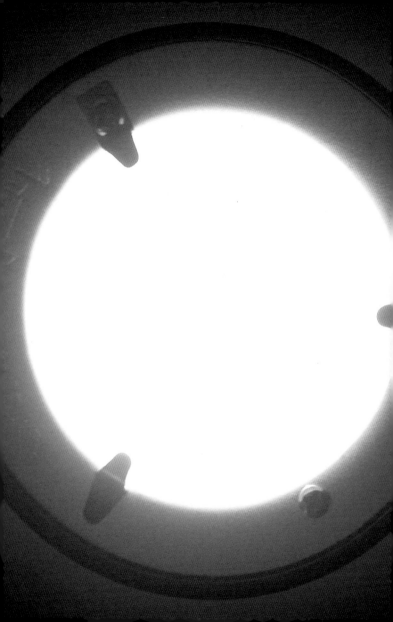

Lighting

Spend nothing – save money

35. If there's nobody in the room, or the room is bright enough without having lights on, switch the lights off. Get into the habit; it costs nothing and is really simple and effective.

36. Use natural light where possible.

37. Beware of 'uplighters': many consume a lot of electricity, using high-wattage bulbs of 300w or greater – that's the equivalent of over 30 low-energy light bulbs! Use energy-efficient spotlights instead.

38. Halogen bulbs consume less electricity than conventional light bulbs, but they generally need to be used in larger numbers because each bulb only lights up a small area, so you may end up using more electricity.

> We spend 10% of our electricity bills on lighting.

39. Have candlelit suppers.

Keep strip lights on, or switch them off?
Some people think that keeping strip lights on is cheaper
and consumes less electricity than switching them on and
off, because to restart these lights uses considerable
electricity. Restarting fluorescent tubes does require some
energy, *but only very little.* If you're going to be out of the
room for more than a couple of minutes, switch them off.

Spend a Little - Save More

40. Use energy-efficient light bulbs, as they last about 12
times longer than ordinary bulbs and consume about ⅕ of
the energy. They come in all shapes and sizes, including
spotlights. Many energy companies and some councils are
even giving them away.

Energy-efficient light bulbs are cheap
to run because they mainly make light
rather than heat. 90% of the energy
used by traditional bulbs is wasted
in producing heat.

Cooking

Cooking

Spend nothing – save money

41. Select the correct saucepan size for the heating element or gas flame.

42. Cut food into small pieces before cooking – it will cook more quickly.

43. Put a lid on top of the pan when you can; your meal will cook much more quickly, and you won't be wasting energy.

44. Use an electric kettle to boil water for cooking.

> **Fan-assisted electric ovens warm up more quickly, distribute the heat more evenly, and use about 20% less electricity than a conventional oven.**

45. Keep your kettle free of limescale – it will be more efficient. Fill it with a mixture of ⅔ water and ⅓ vinegar and leave overnight. Rinse it out well, fill it with water, boil the water and throw it away.

46. If you are cooking with a saucepan, turn down the heat when it comes to the boil. You don't need as much heat to keep a pot boiling as you do to get it to the boil, and the contents will cook just as quickly.

47. Make toast in a toaster rather than under the grill if possible.

48. If you're cooking vegetables in saucepans, only use sufficient water to cover them.

49. Consider using a pressure cooker for cooking some foods – it reduces cooking times dramatically.

> **Our demand for electricity grows by about 3% every year.**

50. 'Slow cookers' are a really cheap way of cooking. The cooker gently simmers away all day, using little more power than a conventional light bulb.

51. If you're cooking a meal in the oven, don't be tempted to keep on opening the oven door to see how it's all going, as you lose a lot of heat doing this.

52. Plan ahead: get ready-made meals out of the freezer early enough for them to defrost without using energy.

53. If you are in a hurry, heat or defrost ready-made meals in a microwave rather than a conventional oven.

54. Don't over-fill an electric kettle: just put in the amount of water you want, but make sure you cover the element. You'll use less energy, it will cost less, and will come to the boil more quickly.

55. When cooking rice, turn off the heat 5 minutes before the end of cooking time, keep the lid on and let it finish cooking in its own steam.

56. Use a steamer for vegetables – you can cook two or three vegetables on one element or gas ring.

57. Make one-pot meals that only need one element or gas ring.

58. Use your oven efficiently by filling up as much of the space as possible.

59. Cook two days' meals at once in the oven and utilise the space. Reheating will use less energy than starting from scratch on day two.

60. When using a non-fan-assisted oven, food will cook more quickly on the top shelf — it is much hotter than the bottom.

61. Where appropriate use the grill rather than the oven.

Spend a little - save more

62. Electric kettles vary in the amount of electricity they consume. When you need to replace yours, choose one with the minimum energy consumption.

> Electric kettles consume surprisingly large amounts of energy because they are used frequently, generally heat more water than is needed, and have to bring the water up to boiling point — an extremely energy-hungry process.

63. Check out microwave ovens. They consume about 80% less electricity than a conventional oven.

64. If you are replacing your electric oven, consider a fan-assisted model as they are cheaper to run.

Keeping things cool

Fresh_Vegetables

Keeping things cool

Spend nothing – save money

65. Wait until hot food has cooled down before putting it into the fridge.

66. Don't keep the fridge door open any longer than necessary.

67. Keep fridges and freezers well away from heat sources such as cookers, dishwashers and washing machines.

68. If possible, site fridges and freezers out of direct sunlight, as your appliance will use more energy trying to keep cool in the sun.

69. Try and keep your fridge and freezer full; they will use less electricity.

> Fridges and freezers are never turned off – although they may not appear to use much energy, in an average home they are responsible for about 1/3 of the total electricity bill.

70. If your freezer isn't full, fill empty spaces with scrunched-up paper or bubble wrap to stop warm air circulating when it is opened.

71. Defrost food by putting it in the fridge the night before you want to use it. This will cool the fridge down and reduce its power consumption.

72. Keep the metal grids (condenser coils) at the back of fridges and freezers clean and dust-free, and not jammed up against the wall; this allows the air to circulate more easily around them, and makes them more efficient.

> A chest freezer uses less electricity than a front-opening model because the cold air doesn't fall out every time the freezer is opened.

73. If you have a fitted kitchen with a built-in fridge or freezer, make sure there is ample ventilation to allow for air circulation around the condenser coils.

74. Defrost the fridge and freezer regularly. If the ice inside gets more than 5mm thick, the appliances become inefficient.

Spend a little – save more

75. Consider buying an energy-efficient freezer to replace older appliances. You should recover the cost remarkably quickly.

76. Check the door seals on your fridge and freezer: shut the door on a £5 note. If you can pull it out easily, or if your seals are damaged, they need replacing.

> A new 'A' energy-rated fridge consumes about ⅓ of the electricity of some of the older models.

Washing and drying clothes

Washing and drying clothes

Spend nothing – save money

77. When washing clothes by hand, there is no need to have the water hot. Most non-greasy dirt will wash out easily with cold water and detergent. Cold water is fine for rinsing your clothes afterwards.

78. Wait until you've got a full load before using your washing machine – using the 'half load' programme does not save you half the energy, water or detergent.

79. Use a lower temperature wash for clothes which aren't very dirty: for most washes, 40°C is just as good as 60°C.

> Washing clothes at 60°C uses almost twice as much energy as a 40°C wash.

80. Use the economy programme where possible.

81. If your machine has a cold wash option, try using it for lightly soiled clothing. Most detergents work extremely well at low temperatures.

82. If possible connect both your hot and cold washing machine hoses to your hot and cold water pipes. This will enable the machine to use readily available hot water rather than having to heat cold water from scratch.

83. If you live in a hard water area, limescale on your washing machine element will reduce its efficiency. Every couple of months get rid of it by running the machine empty on a wash cycle using 200ml of white vinegar in the detergent tray. There are also de-scaling tablets available.

> Energy-efficient washing machines use about ⅓ less electricity than older machines. The savings will more than cover the price of a new machine.

84. Air-dry your clothes on clothes racks or lines if possible – tumble dryers are very energy-hungry appliances.

85. If you have to use a tumble dryer, then spin dry or wring the clothes before putting them in it. Clean out the 'fluff filter' every time you use the dryer: this improves the efficiency and your clothes will dry more quickly.

86. Switch the tumble dryer off when it has finished. It consumes almost 40% of the power whilst on stand-by.

Washing dishes

Washing dishes

Spend nothing – save money

87. When washing dishes by hand, fill a bowl with warm water and a little detergent, washing the 'cleaner' items first. Use cold water for rinsing.

88. If you use a dishwasher, wait until it is full before using it. Don't be tempted by the 'half-load' facility, as it is nowhere near as energy-efficient.

89. Use the 'economy' or 'eco' programme if your dishwasher has one. It will use less electricity and take less time.

90. Switch your dishwasher off completely when it has finished; it is still consuming electricity on stand-by.

91. If you switch off the machine and open the door when the dishwasher enters its 'drying phase', the dishes will dry naturally, saving a considerable amount of energy.

When you buy a new appliance, get an 'A'-rated energy-efficient model; they cost less to run, save you money and contribute less to climate change.

Electrical
appliances
and gadgets

Electrical appliances and gadgets

Our appetite for electrical appliances continues to grow, as does, of course, our need for even greater quantities of electricity to power them. Fridges and TVs have become bigger as mobile phones, computers, iPods and the like have got smaller. We now have electrical appliances in nearly every room of the house.

Spend nothing – save money

92. Turn off the chargers for your mobile phone and laptop when not in use.

93. Turn off TVs, radios, stereos and computers when not in use.

94. Turn your iron off just before you finish ironing, and use the residual heat for the last few clothes.

> At any one time in most households an average of 8 appliances are left on stand-by. In the average UK household the TV is left on stand-by for more than 17 hours a day.

95. Ask the **Energy Saving Trust** do a free energy check on your home. You can reduce your energy consumption, knock hundreds of pounds off your energy bills, and cut down your CO_2 emissions.

96. Use your electricity meter to see for yourself which appliances use the most electricity: have a look at your meter whilst somebody is switching on kettles, toasters, tumble dryers, electric instant showers etc.

> Smart Meters, which give 'live information' about the cost of electricity, gas and water and also store details about your previous usage, are being considered for installation in UK homes.

Spend a little - save more

97. Think about buying a small portable monitor that shows you how much electricity you are using, how much it is costing, and the CO_2 you are adding to the atmosphere.

98. Buy a steam iron: although they use slightly more electricity than dry irons, they are more efficient and take less time.

> Most video recorders and set-top boxes are never turned off. Even in stand-by mode they consume about 85% of the power that they use when working. Some appliances use even more than that.

99. If you are replacing your computer, consider a laptop – they are more energy-efficient.

100. Spread the word!

If you do nothing else . . .

Switch off the lights!

Renewable energy and your home

Renewable energy is energy produced by a source that continually renews itself. Well-known sources are the sun, moving water, wind and plant materials. This energy can be used for space heating and hot water heating, and to produce electricity for your home.

By using renewable energy instead of conventional energy sources, you can reduce the amount of carbon dioxide your household produces. This will reduce your contribution to climate change and save you a considerable amount of money once installed, as most of these energy sources will provide endless free energy, and reduce the impact of gas and electricity price rises.

Providing you have already taken some basic steps to reduce your energy consumption, there will probably be grants available to help you pay for the purchase and installation of a renewable energy system. These grants can be quite substantial. *(See 'Advice and grants' on page 74)*

When considering the purchase and installation of a renewable energy system, you need to consider:

The suitability of your home, such as: Do you have a south-facing roof or wall? Is your house exposed to the wind?

Payback (the amount of time it takes for the renewable energy system to pay for itself). This varies considerably according to which system you install.

Initial cost Some systems are dramatically cheaper than others to buy and install.

The following is a brief outline of renewable energy systems.

Solar power

Energy from the sun can be used both to provide domestic hot water and to produce electricity for your home. Different technologies are used for each.

To produce domestic hot water

Solar heating panels use the sun's energy to heat domestic hot water. This energy typically reduces your water heating bill by 65%–75%.

Solar heating systems work in conjunction with your conventional domestic hot water system.

Most south-facing roofs, walls or gardens are suitable for the installation of solar heating panels.

To produce electricity

Photovoltaic (PV) cells convert sunlight to electricity. This electricity is fed into the mains grid, thereby reducing your electricity bill. The PV cells can be put on a south-facing roof or wall, provided that they are strong enough to support the additional weight and are not shaded by trees or other buildings.

Small-scale wind turbines

To produce electricity

Wind turbines convert moving wind into electricity. For many houses in the UK, a new breed of micro-turbine that attaches to your chimney or roof is the most convenient and practical.

The electricity produced by micro wind turbines is fed back into the mains electricity grid, thereby reducing your electricity bill. Your house needs to be exposed to the wind to make this system suitable for you.

Biomass (biofuels)

To heat your house and hot water

Biomass or biofuels are materials such as wood or straw which grow quickly and can be burnt to release heat, for space heating and domestic hot water. Biomass is different from all the other renewable energy sources because the fuel generally has to be purchased.

Biomass is a renewable energy source because:

- The materials are quick to grow, absorbing CO_2 in the process.
- The CO_2 released when it is burned balances that which was absorbed during the growth of the material, effectively making the process carbon-neutral.

Wood (in the form of logs or pellets) is the most commonly used biofuel. It should be burnt in an efficient, controllable manner, either in stand-alone stoves or in boilers.

Ground source heat pumps

To heat your house and hot water

Heat pumps take heat from several metres under the ground (which remains at about $12°C$ all year round) and use it to heat your house – just like a refrigerator in reverse. They can also be used to warm water before it enters your domestic hot water heater, thereby saving on energy used. If you want to install a heat pump, you will need sufficient space outside to dig either a trench or a borehole.

Although heat pumps are run by electricity, they are very efficient: for every unit of electricity used to run the heat pump, about four units of heat energy are created.

Small-scale hydro power

To produce electricity

If you are fortunate enough to have a fast-moving stream or river running near your house, it might be possible to generate electricity from the moving water. Though not the simplest of renewable energy systems to install, hydro schemes have the capacity to generate substantial amounts of electricity, which can then be sold back to your electricity company.

The potential source of power will need to be assessed initially before any other steps are taken. Costs of hydro power schemes vary hugely according to the size of the project, but hydro schemes can sometimes offer high returns.

If you think your local river has the potential to generate electricity, consider forming a community hydro project. There are people operating successful schemes who are willing to provide advice.

Advice and grants

There are many organisations that exist just to help you save money. Grants and advice are available to help you:

- Insulate your home
- Generate your own electricity and heating from renewable sources
- Improve your heating
- Purchase energy saving appliances
- Generate electricity and heat for your community from renewable sources

Grants are available from the government, your energy supplier and your local authority, and can be quite substantial, especially if you are considering installing a community scheme to generate electricity or heat from a renewable source.

As well as grants, some councils now offer a considerable reduction in council tax if you have insulated cavity walls. If you are in receipt of benefits, you may be able to obtain free cavity wall and loft insulation for your home.

For advice, grants and offers available in your region contact the **Energy Saving Trust** (see page 77 for details). Their web site is very user-friendly and has loads of helpful information.

Example of an energy label

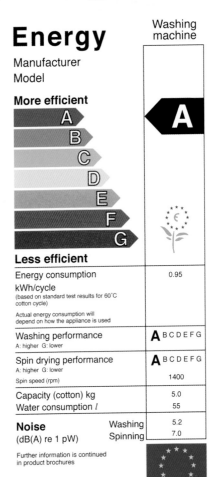

Energy

Manufacturer
Model

More efficient

A

B

C

D

E

F

G

Less efficient

	Washing machine
	A
Energy consumption kWh/cycle (based on standard test results for 60°C cotton cycle) Actual energy consumption will depend on how the appliance is used	0.95
Washing performance A: higher G: lower	**A** B C D E F G
Spin drying performance A: higher G: lower Spin speed (rpm)	**A** B C D E F G 1400
Capacity (cotton) kg Water consumption *l*	5.0 55
Noise (dB(A) re 1 pW) Washing Spinning	5.2 7.0

Further information is continued
in product brochures

Resources

DirectGov

The website of the UK government. Use it to find the contact details of your local authority. Contact them to see if what grants or special offers are available.

Website: www.direct.gov.uk

Energy Future

A web site that aims to clear away the confusion and dispel the myths and misconceptions about topics such as climate change and DIY energy – and more.

Website: www.energyfuture.org.uk

The Energy Retail Association

The ERA represents the six major electricity and gas suppliers in Britain's competitive energy market.

Website: www.energy-retail.org.uk

The Energy Saving Trust

The EST provides independent advice and information about energy efficiency, insulation and renewable energy options, and the availability of grants. These are the first people to ring to find out what help, advice, grants and offers are available in your area.

Phone: 0800 512 012
Website: www.est.org.uk/myhome/gid

The Energy Supply Ombudsman

An independent body which rules on any disputes between householders and energy supply companies on issues relating to bills and switching energy suppliers.

Phone: 0845 055 0760
Website: www.energy-ombudsman.org.uk

Energywatch

This independent gas and electricity watchdog provides free and impartial advice to get energy consumers the best deal and take up complaints on their behalf.

Phone: 08459 06 07 08
Website: www.energywatch.org.uk

Home Heat Helpline

A free service for anyone worried about paying their energy bill. Expert advisors can help you with claiming grants, advise on saving money on energy bills, and on accessing company and government benefits. Funded by the Energy Retail Association.

Phone: 0800 336699
Website: www.homeheathelpline.org

The National Energy Foundation

An independent educational charity working for a more efficient, innovative and safe use of energy, and to increase the public awareness of energy in all its aspects. Currently it is working in the areas of renewable energy and energy efficiency.

Phone: 01908 665555
Website: www.nef.org.uk

Ofgem

The regulator for Britain's gas and electricity industries. Its role is to promote choice and value for all customers.

Website: www.ofgem.gov.uk

Other Green Books Guides

Water: use less – save more
by Jon Clift and Amanda Cuthbert
100 water-saving tips for the home, in full colour.

Reduce, Reuse, Recycle: an easy household guide
by Nicky Scott
An easy-to-use A–Z household guide to recycling.

Composting: an easy household guide
by Nicky Scott
Tells you everything you need to know for successful home composting.

Ecology Begins at Home
by Archie Duncanson
A personal account of living a low-impact lifestyle.

About the authors

Jon Clift has a Masters degree in Sustainable Environmental Management. He works as a freelance environmental consultant and lives in Salcombe, South Devon.

Amanda Cuthbert is the author of four books, including (with Jon Clift) *Water: Use less, save more*. She now works as a freelance writer and marketing consultant and lives in Devon.